梁燕 編著

新手入廚系列

炒出美味

前言

即叫即炒、熱辣辣的「大排檔」美食是香港人至愛的集體回憶之一，廚師把食材放在鑊中炒一炒、拋兩拋，再贊一贊酒，在分秒之間，食物的味道得以昇華，一道道鑊氣十足的小菜立即上碟。除了小菜的色、香、味，最重要是將熱騰騰的菜式送到桌上。

用快炒的方式將食物快速炒熟，用油量較少，可保留食物中的營養素和食材的原有味道。但要注意用大火炒餸，食油溫度容易過熱，食油也有可能產生化學變化。

讀者家中爐頭的火力雖然不及大牌檔的火水爐或酒樓、餐廳的石油氣爐，但在家中也可練成幾道「鑊氣」撚手小菜如薑葱炒蟹、雪菜炒年糕、乾隆炒鴿鬆、炒雜錦粒粒和回鍋肉等，才稱得上懂得享受人生。

目錄

雞鴨類

豬牛類

看圖買材料做菜

Buy ingredients according to the pictures

牛肉（炒）
要帶少許肥肉。

Beef (for stir-frying)
Should contains a little fat.

豬肉（炒）
要購買柳梅，因比較稔。

Pork (for stir-frying)
Purchase pork fillet as the texture is more tender.

魚柳
顏色鮮明。

Fish fillets
Bright colors.

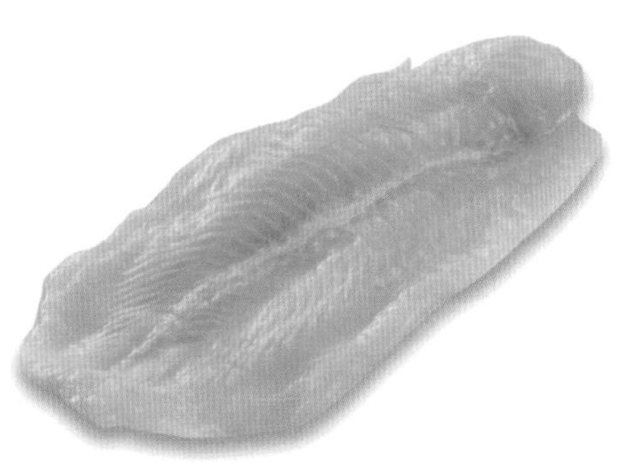

蜆

蜆殼開了不合的，即表示死了，不可購買。

Clams

Should not buy clams with shell opened as that means the clam is dead.

蟹

拍拍蟹蓋，蟹的眼睛活動得很靈活表示未死。

Crabs

Pat the shell of crab, the crab is alive if the eyes are very flexible.

蝦

有光澤，蝦頭不要呈黑色。

Shrimps

Shrimps should look shiny. Do not buy those with black heads.

生魚

身上沒有瘀紅或傷痕。

Raw fish

Do not have any red or stasis wound on fish body.

洋葱

完整光滑，沒有損傷。

Onion

Complete, smooth and not rotten.

蜜糖豆

青綠而脆。

Sugar snap peas

Green and crispy.

番茄

紅色圓大。

Tomatoes

Round, big and red in color.

韭黃

黃色，尾部不要變黑和爛。

Chives

Yellow in color, tail part should not be black and rotten.

沙葛

表面沒有損爛。

Shage

Surface is not rotten.

年糕

顏色要白，不要有黑點。

Pudding cake

Color should be white and do not have black spots.

鹹酸菜

不要選擇太黃的，可能加了色素。

Pickled cabbage

Do not choose those in dark yellow color, pigment may be added.

XO 醬

成份要多瑤柱。

XO paste

Should contains more dried scallops.

做菜的常用語

Common phase of cooking

滾刀塊

切成不規則的形狀。

Pieces in rolling cut

Cut ingredients into irregular shapes by rolling the ingredients.

起雙飛

魚肉切第一刀時不要切斷，要餘下少許，再切第二刀就要切斷，兩片魚肉都要薄而相連。

Butterflying

Do not cut off fish at the first cut, then cut off at the second cut, two slices should be thin and connected continuously.

勾芡（埋芡）

食物煮熟後加生粉水。

Thickening

Ingredients are cooked and add cornstach solution to thicken.

玻璃芡

生粉加水不加其他顏色。

Transparent thickening sauce

Cornstarch with water without adding other colors.

汆水

食物放滾水中再撈起。

Blanch

Blanch ingredients in boiling water and then dish up.

過冷河

食物放滾水中再撈起，放凍水中沖過，瀝乾水分。

Cooling off in water

Blanch ingredients in boiling water and then dish up, put in cold water and drain.

泡油

將食物放滾油中一會即撈起。

Blanch with oil

Place food in boiling oil for a while and drain.

回鑊

食物泡油後再放回鑊中與其他材料一起炒。

Return to wok

Blanch food in boiling oil and then return to wok and stir-fry along with other ingredients.

買回來的材料怎麼辦？

劏蟹 Gut Crabs

洗蟹：用牙刷擦去腳邊的污泥，但要小心紮着的繩不要鬆脫。

Clean crabs: Brush the feet of crabs with a toothbrush but do not loose the rope holding crabs.

劏蟹步驟見下圖：

See pictures below for cutting crabs:

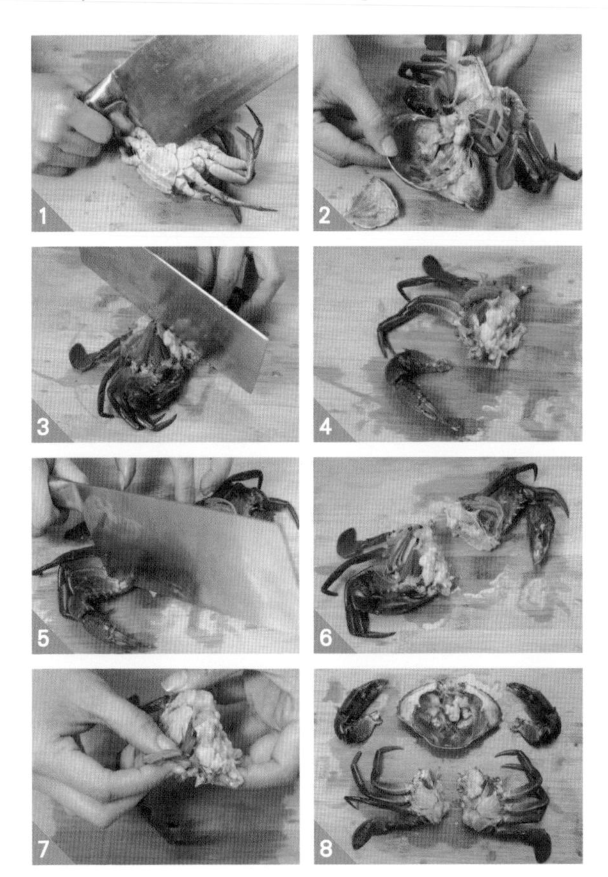

切牛肉 Cut Beef

牛肉用橫紋切才不會韌。

Cut beef against the stripes to avoid tough texture.

切魚片 Cut Fish Fillets

魚肉切第一刀時不要切斷，要餘下少許，再切第二刀就要切斷，兩片魚肉都要薄而相連。

Do not cut off fish at the first cut, then cut off at the second cut, two slices should be thin and connected continuously.

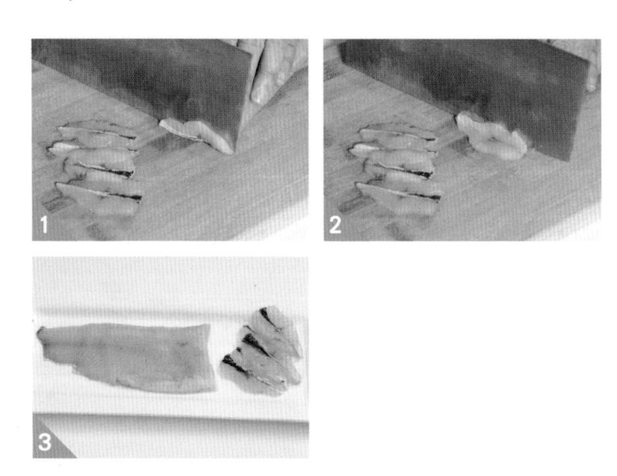

買回來的材料怎樣處理？

How to deal with the ingredientss bought ?

魚：買回來要即時放入雪櫃保持新鮮。

Fish: Put fish into refrigerator immediately after buying to keep fresh.

蝦：去殼去腸，洗淨，用少許鹽水浸着，放入雪櫃，可令蝦爽脆。

Shrimp: Shell and devein shrimp, rinse and soak with some salty water. Put in refrigerator can keep texture crispy.

菜：買回來要保持新鮮，用報紙包好，放入雪櫃。

Vegetables: To maintain vegetables fresh, wrap with newspaper and place in refrigerator.

辣椒：放冰箱內可保存數個月。

Chili: Red chili can be stored in refrigerator for a few months.

炒餸也好味

Stir-fry

原理：以油作為傳熱媒介，將加工成的食材用旺火快速加熱，充分攪拌，使油、食材和調味料迅速融合而成菜。

Principle: Oil is the heat transfer medium: heat the processed ingredients over high heat quickly. Stir-fry thoroughly to integrate oil, ingredients and seasonings into a dish.

優點：保持食材的鮮味；調味方法變化多，味覺和視覺上豐富多彩。

Advantages: To maintain freshness of ingredients. Variety in seasonings, taste and visual effect is rich.

20 分鐘
20 minutes

4 人
Serves 4

時蔬炒魚塊

Stir-fried Fish Fillets with Vegetables

材料 | Ingredients

菜芯 200 克	200 g Choi Sum
魚柳 2 條	2 slices fish fillets
薑 2 片	2 slices ginger
米酒 2 茶匙	2 tsps rice wine

醃料 | Marinade

雞蛋白 1/2 隻	1/2 egg white
酒 1 茶匙	1 tsp wine
生粉 1 茶匙	1 tsp cornstarch
胡椒粉 1/2 茶匙	1/2 tsp pepper
雞粉 1/2 茶匙	1/2 tsp chicken powder

調味料 | Seasonings

鹽 1/2 茶匙	1/2 tsp salt
糖 1/3 茶匙	1/3 tsp sugar

做法 | Method

1. 魚柳洗淨，切成魚塊，瀝乾水分，加入醃料拌勻，醃約 15 分鐘。
2. 菜芯棄掉老葉，洗淨，切段。
3. 燒熱鑊，下油約 1 碗，待油燒滾後下魚塊泡油至 7 成熟，取出並用廚房紙稍吸去油分。
4. 再燒熱鑊，下油約 1/2 湯匙，爆香薑片，下菜芯略炒，加調味料拌勻，將魚塊回鑊炒勻，灒酒，即可上碟。

1. Rinse fish fillets, cut into pieces, drain. Marinate for about 15 minutes.
2. Discard old leaves of Choi Sum, rinse and cut into sections.
3. Heat wok and bring a bowl of oil to a boil, add fish fillets and cook until 70% done. Dish up and absorb excess oil by kitchen paper.
4. Heat wok again, add 1/2 tbsp of oil and sauté ginger slices. Sauté Choi Sum, add seasonings and mix well. Return fish fillets to wok, stir well, drizzle wine. Serve.

入廚貼士 | Cooking Tips

- 魚柳加酒醃可除去魚的腥味。
- Marinate fish fillets with wine can remove unpleasant smell.

鹹酸菜炒魚鬆

Stir-fried Minced Dace with Pickled Cabbage

材料 | Ingredients

原味絞鯪魚肉 200 克	200 g minced dace (original flavor)
鹹酸菜梗 2 塊	2 stalks pickled cabbage
葱段 2 棵	2 stalks spring onion (sectioned)
紅燈籠椒 1 隻	1 red bell pepper
蒜蓉 1 茶匙	1 tsp minced garlic

醃料 | Marinade

胡椒粉 1/2 茶匙	1/2 tsp pepper
鹽 1/2 茶匙	1/2 tsp salt
生粉 1 1/2 茶匙	1 1/2 tsps cornstarch
水 2 湯匙	2 tbsps water

調味料 | Seasonings

生抽 2 茶匙	2 tsps soy sauce
糖 2 茶匙	2 tsps sugar
生粉 1 茶匙	1 tsp cornstarch
麻油 1/2 茶匙	1/2 tsp sesame oil
胡椒粉 1/2 茶匙	1/2 tsp pepper

做法 | Method

1. 鹹酸菜梗洗淨，切絲，用鹽水浸 15 分鐘，榨乾水分待用。
2. 紋鯪魚肉加醃料醃 15 分鐘。
3 . 紅燈籠椒洗淨，去籽，切絲。
4. 燒熱鑊，下油約 1/2 湯匙，下紋鯪魚肉，壓成圓餅形，煎至兩面金黃，待涼後切成條狀。
5. 燒熱鑊，下油約 1/2 湯匙，爆香蒜蓉，下紅燈籠椒、鹹酸菜，紋鯪魚肉條，加入調味料和葱段拌勻即可。

1. Rinse pickled cabbage stems and shred. Soak in salty water for 15 minutes, drain and squeeze out excess water.
2. Marinate minced dace for 15 minutes.
3. Rinse red bell pepper, seed and shred.
4. Heat wok and add 1/2 tbsp of oil, pan-fry minced dace and press into a round pie. Pan-fry until golden on both sides, leave to cool and cut into strips.
5. Heat wok with 1/2 tbsp of oil, sauté minced garlic, red bell pepper, pickled cabbage stems and stripped minced dace. Add seasonings and spring onion and mix well.

入廚貼士 | Cooking Tips

- 鹹酸菜用鹽水略浸可減去鹹味。
- The salty taste of pickled cabbage could be reduced if soaked in salty water.

20 分鐘
20 minutes

4 人
Serves 4

勝瓜雲耳炒生魚片

Stir-fried Snakehead with Angled Luffa and Ear Fungus

材料 | Ingredients

勝瓜（絲瓜）2 條
雪耳 4 朵
生魚（去骨）1 條
蒜蓉 2 茶匙
薑 2 片

2 angled luffa
4 ear fungus
1 snakehead fish (deboned)
2 tsps garlic
2 slices ginger

醃料 | Marinade

鹽 1/4 茶匙	1/4 tsp salt
胡椒粉少許	Pinch of pepper

調味料 | Seasonings

鹽 1/2 茶匙	1/2 tsp salt
雞粉 1/2 茶匙	1/2 tsp chicken powder
生抽 1/2 茶匙	1/2 tsp soy sauce
糖 1/4 茶匙	1/4 tsp sugar
水 1/2 杯	1/2 cup water

做法 | Method

1. 勝瓜去皮，洗淨，切滾刀塊。
2. 雪耳浸發後洗淨，去蒂，切小朵。
3. 生魚洗淨，切雙飛，下醃料拌勻。
4. 燒熱鑊，下油約 1/2 湯匙，爆香蒜蓉和薑片，下勝瓜和雪耳炒至熟，加入調味料拌勻。加入生魚片，輕力炒至熟透即可。

1. Peel angled luffa, rinse and cut into irregular pieces.
2. Soak ear fungus until soft, rinse and remove stalks, cut into small pieces.
3. Rinse snakehead, butterfly, marinate and mix well.
4. Heat wok with 1/2 tbsp of oil, sauté minced garlic and ginger slices, add angled luffa and ear fungus and stir-fry until done. Add seasonings and mix well. Add snakehead slices and gently stir-fry until done.

入廚貼士 | Cooking Tips

- 生魚可請魚檔代為去骨起肉。
- May ask the staff help to remove the bones of snakehead.

珊瑚蚌炒蜜糖豆

Stir-fried Canadian Red Sea Cucumber with Sugar Snap Peas

材料 | Ingredients

珊瑚蚌 180 克
中蝦 8 隻
蜜糖豆 150 克
芹菜 2 棵
乾葱蓉 2 茶匙
薑米 1 茶匙
米酒 2 茶匙

180 g Canadian red sea cucumber meat
8 medium shrimps
150 g sugar snap peas
2 stalks Chinese celery
2 tsps chopped shallots
1 tsp minced ginger
2 tsps rice wine

25 分鐘
25 minutes

4 人
Serves 4

醃料 | Marinade

胡椒粉 2 茶匙	2 tsps pepper
生粉 1/2 茶匙	1/2 tsp cornstarch
雞粉 1/4 茶匙	1/4 tsp chicken powder

芡汁 | Thickening

蠔油 1 1/2 湯匙	1 1/2 tbsps oyster sauce
生粉 2 茶匙	2 tsps cornstarch
麻油 1/2 茶匙	1/2 tsp sesame oil
雞粉 1/2 茶匙	1/2 tsp chicken powder
水 3 湯匙	3 tbsps water

做法 | Method

1. 珊瑚蚌洗淨，瀝乾水分備用；蜜糖豆洗淨，摘去兩邊根莖，汆水備用；芹菜洗淨，切去根部和葉，切段。
2. 中蝦洗淨，去殼去腸，放水龍頭下沖 10 分鐘，瀝乾水分。加醃料拌勻，放雪櫃中 1/2 小時。
3. 燒熱鑊，下油約 1/2 湯匙，爆香薑米和乾葱蓉，加入中蝦和珊瑚蚌略炒，濳酒，炒至香味溢出。
4. 芡汁放碗中拌勻，下芡汁煮滚，再放蜜糖豆和芹菜，炒至芡汁稍收乾即可上碟。

1. Rinse Canadian red sea cucumber meat, drain for later use. Rinse sugar snap peas, remove woody outer fibers on both sides, blanch. Rinse Chinese celery, cut roots and leaves, section.
2. Rinse shrimps, shell and devein. Rinse under the tap for 10 minutes and drain. Add marinade and mix well, put in refrigerator for 1/2 hours.
3. Heat wok with 1/2 tbsp of oil, sauté minced ginger and shallots, add shrimps and Canadian red sea cucumber meat. Drizzle wine and stir-fry until aroma comes out.
4. Mix sauce in a bowl. Heat wok and bring thickening to a boil. Add sugar snap peas and Chinese celery, stir-fry until sauce is slightly dried. Serve.

入廚貼士 | Cooking Tips

- 蜜糖豆要汆水以去除草青味，炒時不要炒得太腍，味道才好。
- Blanch sugar snap peas to remove grass flavor, do not overcook to keep crispy texture.

15 分鐘
15 minutes

6 人
Serves 6

象拔蚌炒西蘭花

Stir-fried Geoduck Clams with Broccoli

材料 | Ingredients

西蘭花 1 棵（約 250 克）	1 pc broccoli (around 250g)
象拔蚌 200 克	200 g geoduck clams
蒜蓉 1 茶匙	1 tsp minced garlic
薑汁 1 茶匙	1 tsp ginger sauce

芡汁 | Thickening

清雞湯 3 湯匙	3 tbsps chicken broth
生抽 2 茶匙	2 tsps light soy sauce
生粉 1 茶匙	1 tsp caltrop starch
糖 1/2 茶匙	1/2 tsp sugar
麻油少許	Some sesame oil

做法 | Method

1. 象拔蚌洗淨，切薄片。
2. 西蘭花洗淨，切細。
3. 燒熱一鍋水，下 1 湯匙油，放入西蘭花汆水，瀝乾。
4. 燒熱 2 湯匙油，爆香蒜蓉，下象拔蚌快手兜炒，再加入薑汁和西蘭花，下芡汁炒勻便可上碟。

1. Rinse and cut geoduck clams into thin slices.
2. Rinse broccoli and cut into small pieces.
3. Heat a wok of water, add 1 tbsp of oil, blanch broccoli and drain.
4. Heat 2 tbsps of oil, sauté minced garlic and toss with geoduck clams quickly. Stir in ginger sauce and broccoli, then toss with thickening until well combined and dish up.

入廚貼士 | Cooking Tips

- 這款快炒方法，簡單方便又容易處理，但是必須緊記即炒即吃。
- This stir-frying method is simple and easy to handle but remember that the dish must be served hot.

XO 醬炒蝦仁

Stir-fried Shrimps with XO Paste

材料 | Ingredients

中蝦 250 克
青燈籠椒 1 隻
紅燈籠椒 1 隻
蒜頭 3 粒
酒適量

250 g medium shrimps
1 green bell pepper
1 red bell pepper
3 cloves garlic
Some wine

20 分鐘
20 minutes

4 人
Serves 4

醃料 | Marinade

雞蛋白 1/2 隻	1/2 egg white
胡椒粉 1/2 茶匙	1/2 tsp pepper
鹽 1/2 茶匙	1/2 tsp salt

汁料 | Sauce

蠔油 2 茶匙	2 tsps oyster sauce
XO 醬 1 茶匙	1 tsp XO paste
糖 1/2 茶匙	1/2 tsp sugar
麻油 1/2 茶匙	1/2 tsp sesame oil

做法 | Method

1. 蝦去殼去腸，洗淨，在背後剪開，加入醃料拌勻。
2. 蒜頭去衣，洗淨、剁蓉。
3. 青、紅燈籠椒洗淨，去籽，切滾刀塊。
4. 燒熱鑊，下油約 1/2 湯匙，爆香蒜蓉，下青、紅燈籠椒炒勻，再加入汁料，灒酒，下蝦仁炒至熟即可上碟。

1. Shell and deveine shrimps, rinse. Cut at the back, add marinade and mix well.
2. Peel garlic, rinse and chop finely.
3. Rinse green and red bell peppers, seed and cut into irregular shapes.
4. Heat wok with about 1/2 tbsp of oil, sauté minced garlic, add green and red bell peppers and stir well. Then add sauce and drizzle wine, stir-fry until shrimps done. Serve.

入廚貼士 | Cooking Tips

- 蝦洗淨後放水龍頭下沖水 15 分鐘，用廚房紙抹乾水分才加醃料，再放雪櫃中冷藏 1 小時，可令蝦肉爽脆。
- Place shrimps under running tap for 15 minutes. Wipe dry with kitchen paper before adding marinade. Put in refrigerator for 1 hour to make shrimps crispy.

15 分鐘
15 minutes

4 人
Serves 4

蝦仁炒滑蛋
Scrambled Eggs with Shrimps

材料 | Ingredients

蝦仁 220 克
雞蛋 6 隻

220 g shelled shrimps
6 eggs

醃料 | Marinade

雞蛋白 1/2 隻	1/2 egg white
鹽 1/2 茶匙	1/2 tsp salt
胡椒粉 1/2 茶匙	1/2 tsp pepper
生粉 1/2 茶匙	1/2 tsp cornstarch

調味料 | Seasonings

鹽 1/2 茶匙	1/2 tsp salt
生粉少許	Some cornstarch

做法 | Method

1. 蝦仁去腸，洗淨後抹乾水分，加入醃料略醃。
2. 燒熱鑊，下油約 1 湯匙，用大火炒蝦仁至八成熟，盛起待涼。
3. 將雞蛋打入大碗中，拌勻成蛋液。
4. 將蝦仁放入打好的蛋液中，加入調味料拌勻。
5. 燒熱油鑊，下油約 1 湯匙，將蝦仁蛋液倒入，快速炒至蛋液凝固，盛起。

1. Devein shrimps, rinse and wipe dry. Add marinade and mix well.
2. Heat wok with 1 tbsp of oil, stir-fry shrimps until 80% done. Dish up and leave to cool.
3. Whisk eggs into a large bowl.
4. Add shrimps into beaten eggs, add seasonings and mix well.
5. Heat wok with 1 tbsp of oil, add whisked eggs and shrimps, quickly stir-fry until egg solidifies, dish up.

入廚貼士 | Cooking Tips

- 將雞蛋打勻後加入少許生粉，可以令蝦仁炒蛋減少釋出水分。
- Add some cornstarch when beating eggs could prevent scrambled eggs from squeezing excess water.

沙拉蝦球

Stir-fried Shrimps in Salad Dressing

材料 | Ingredients

中蝦 400 克
青椒 1 個
沙律醬 2 湯匙
麵粉 2 茶匙

400 g medium shrimps
1 green pepper
2 tbsps salad dressing
2 tsps flour

25 分鐘
25 minutes

4 人
Serves 4

醃料 | Marinade

雞蛋白 1 隻	1 egg white
生粉 1 1/2 茶匙	1 1/2 tsps cornstarch
胡椒粉 1/2 茶匙	1/2 tsp pepper
鹽 1/4 茶匙	1/4 tsp salt

做法 | Method

1. 中蝦去殼，背部剖開，去腸，洗淨，瀝乾水分。以醃料醃 10 分鐘。
2. 青椒洗淨，切半，去籽，切滾刀塊。
3. 麵粉放碟中，將中蝦撲上麵粉。
4. 燒熱鑊，下油約 1 碗，待油燒滾後下中蝦炸熟，取出並用厨房紙稍吸去油分。
5. 再燒熱鑊，下油約 1/2 湯匙，將青椒和蝦回鑊拌勻，最後加入沙律醬少許炒勻，上碟。

1. Shell shrimps, slit the back and remove intestines, rinse and drain. Marinate for 10 minutes.
2. Rinse green peppers, cut in halves, seed and cut into irregular pieces.
3. Put flour onto a plate, pounce shrimps with flour.
4. Heat wok and bring a bowl of oil to the boil, deep-fry shrimps until done. Dish up and absorb oil by kitchen paper.
5. Heat wok with 1/2 tbsp of oil again, return green pepper and shrimps to wok and mix well. Finally add some salad dressing and stir well. Serve.

入廚貼士 | Cooking Tips

- 炒好蝦後，可以不加沙律醬，待上碟才擠上沙律醬。
- Another option is add salad dressing after dishing up shrimps.

30 分鐘
30 minutes

4 人
Serves 4

薑葱炒蟹
Stir-fried Crabs with Ginger and Spring Onion

材料 | Ingredients

肉蟹 2 隻	2 crabs
薑 10 片	10 slices ginger
葱 10 棵	10 stalks spring onions
蒜頭 5 粒	5 cloves garlic
生粉 2 湯匙	2 tbsps cornstarch
米酒 2 茶匙	2 tsps rice wine

調味料 | Seasonings

胡椒粉 1 茶匙	1 tsp pepper
鹽 1/2 茶匙	1/2 tsp salt
糖 1/3 茶匙	1/3 tsp sugar

做法 | Method

1. 肉蟹劏好，洗淨，切件，瀝乾水分。
2. 葱洗淨，切去根部和尾部，切段。
3. 生粉放碟中，將肉蟹撲上生粉。
4. 燒熱鑊，下油約 1 碗，待油燒滾後下肉蟹泡油至 7 成熟，取出並用廚房紙稍吸去油分。
5. 再燒熱鑊，下油約 1/2 湯匙，爆香薑片和葱段，將蟹回鑊，加調味料，灒酒，拌勻即可上碟。

1. Rinse and gut crabs, cut into pieces and drain.
2. Rinse spring onion, cut roots and tails, cut into sections.
3. Put cornstarch onto a plate, pounce crabs with cornstarch.
4. Heat wok and bring a cup of oil to the boil, add crabs and cook until 70% done. Dish up and absorb excess oil by kitchen paper.
5. Heat wok with 1/2 tbsp of oil again, sauté ginger slices and spring onion sections. Return crabs to wok, add seasonings and drizzle wine, mix well and dish up.

入廚貼士 | Cooking Tips

- 肉蟹撲上生粉才泡油，可鎖住蟹的肉汁。
- Gravy of crabs could be reserved if they are pounced with cornstarch before blanching in oil.

椒鹽蝦

Salt and Pepper Shrimps

材料 | Ingredients

中蝦 450 克
辣椒 1 隻
蒜蓉 3 湯匙
葱粒 2 湯匙
薑蓉 2 茶匙

450 g shrimps
1 red chili
3 tbsps minced garlic
2 tbsps chopped spring onion
2 tsps minced ginger

30 分鐘
30 minutes

4 人
Serves 4

海產類 Seafood

醃料 | Marinade

生粉 2 茶匙	2 tsps cornstarch
雞粉 1 茶匙	1 tsp chicken powder
鹽 1/2 茶匙	1/2 tsp salt

調味料 | Seasonings

淮鹽 1/2 茶匙	2 tsp spiced salt

做法 | Method

1. 中蝦去殼，背部剖開去腸，洗淨，瀝乾水分，以醃料醃 10 分鐘。
2. 辣椒洗淨，去籽，切碎。
3. 麵粉放碟中，將中蝦撲上麵粉。
4. 燒熱鑊，下油約 1 湯匙，待油燒滾後下中蝦煎至金黃盛起。
5. 再燒熱鑊，下油約 1/2 湯匙，爆香蒜蓉、辣椒蓉和薑蓉，將中蝦回鑊，下調味料拌勻，撒上葱粒即可上碟。

1. Shell shrimps, slit at back and remove intestines, rinse and drain. Marinate for 10 minutes.
2. Rinse red chili, seed and chop.
3. Put flour onto a plate, pounce shrimps with flour.
4. Heat wok and bring 1 tbsp of oil to a boil, pan-fry shrimps until golden. Dish up.
5. Heat wok again with 1/2 tbsp of oil, sauté minced garlic, chili and ginger. Return shrimps to wok, add seasonings and mix well, sprinkle chopped spring onion. Serve.

入廚貼士 | Cooking Tips

- 淮鹽可以自己做：以白鑊炒香鹽和五香粉即成淮鹽。
- Pepper salt can be DIY by stir-frying salt and pepper without adding oil until fragrant.

30 分鐘
30 minutes

4 人
Serves 4

胡椒炒蟹
Stir-fried Crabs with Pepper

材料 | Ingredients

肉蟹 2 隻	2 crabs
薑片 120 克	120 g ginger slices
葱段 100 克	100 g spring onion sections
鮮胡椒 50 克	50 g fresh pepper
葱頭 4 粒	4 cloves shallots
蒜頭 4 粒	4 cloves garlic
辣椒 3 隻	3 red chili
米酒 2 茶匙	2 tsps rice wine

醃料 | Marinade

魚露 1/2 茶匙	1/2 tsp fish sauce

調味料 | Seasonings

糖 1 茶匙	1 tsp sugar
鹽 1/2 茶匙	1/2 tsp salt

做法 | Method

1. 肉蟹劏好，洗淨，切件，瀝乾水分，加入醃料醃 10 分鐘。
2. 葱頭、蒜頭洗淨，去衣，用刀略拍。辣椒洗淨，去籽，切碎。
3. 麵粉放碟中， 將肉蟹撲上生粉。
4. 燒熱鑊，下油約 1 碗，待油燒滾後下肉蟹泡油至 5 成熟，取出並用厨房紙稍吸去油分。
5. 再燒熱鑊，下油約 1/2 湯匙，爆香薑片、葱段、葱頭、蒜頭、辣椒和鮮胡椒。將肉蟹回鑊，炒至香味溢出，灒酒，加入調味料拌匀，蓋鑊蓋焗 5 分鐘。加入下葱段拌匀即可上碟。

1. Gut crabs, rinse and cut into pieces and drain. Marinate for 10 minutes.
2. Rinse shallots and garlic, peel and pat with a chopper. Rinse red chili, seed and chop.
3. Put cornstarch onto a plate, pounce crabs with cornstarch.
4. Heat wok and bring a cup of oil to a boil, add crabs and cook until 50% done. Dish up and absorb excess oil by kitchen paper.
5. Heat wok again with 1/2 tbsp of oil, sauté ginger, spring onion, shallots, garlic, chili and fresh pepper. Return crabs to wok, stir-fry until aroma comes out. Drizzle wine, add seasonings and mix well, cover wok and cook for 5 minutes. Add spring onion sections and mix well. Serve.

入廚貼士 | Cooking Tips

- 白鑊炒香胡椒粒，將其中 1/3 壓碎，味道會比較濃郁。
- Sauté pepper in wok without adding oil, then crush 1/3 of pepper, the taste would be rich.

韭菜沙葛炒蜆米

Stir-fried Mini Clams with Leeks and Shage

25 分鐘
25 minutes

4 人
Serves 4

材料 | Ingredients

沙葛 1/2 個	1/2 shage
韮菜 200 克	200 g leeks
蜆米 150 克	150 g mini clams
蒜蓉 1 湯匙	1 tbsp minced garlic

調味料 | Seasonings

鹽 1/2 茶匙	1/2 tsp salt
胡椒粉 1/2 茶匙	1/2 tsp pepper
糖 1/3 茶匙	1/3 tsp sugar

做法 | Method

1. 韮菜洗淨，切去老梗，切段。
2. 沙葛去皮，洗淨，切條。
3. 蜆米洗淨，瀝乾水分。
4. 燒熱鑊，下油約 1/2 湯匙，爆香蒜蓉，加入沙葛和蜆米略炒，下調味料和少許水拌匀，蓋好焗片刻。最後加入韮菜略炒匀即可。

1. Rinse leeks, cut old stems and section.
2. Peel shage, rinse and cut into strips.
3. Rinse mini clams and drain.
4. Heat wok with 1/2 tbsp of oil, sauté minced garlic. Stir-fry shage and mini clams. Add seasonings and some water, mix well. Cover and cook for a while. Finally, add leeks and stir well.

入廚貼士 | Cooking Tips

- 沙葛有季節性，如買不到可改用豆腐乾或馬蹄。
- Shage is a seasonal ingredient, it could be replaced by dried beancurd or water chestnuts.

25 分鐘
25 minutes

4 人
Serves 4

豉椒炒鮮魷

Stir-fried Squid in Black Bean Sauce

材料 | Ingredients

鮮魷魚 500 克	500 g fresh squid
西芹 100 克	100 g celery
青椒 1 個	1 green pepper
紅燈籠椒 1 個	1 red bell pepper
乾葱 4 粒	4 shallots
蒜蓉 2 茶匙	2 tsps minced garlic
米酒 2 茶匙	2 tsps rice wine
豆豉 1 湯匙	1 tbsp fermented black beans

汁料 | Sauce

蠔油 2 茶匙	2 tsps oyster sauce
麻油 1/2 茶匙	1/2 tsp sesame oil
生粉 1/2 茶匙	1/2 tsp cornstarch
糖 1/4 茶匙	1/4 tsp sugar
水 4 湯匙	4 tbsps water

做法 | Method

1. 鮮魷魚洗淨，撕去薄膜和除去內臟，剞花。
2. 西芹洗淨，撕去老根。
3. 青椒和紅燈籠椒洗淨，去籽，切件。乾葱洗淨，去衣，用刀略拍。
4. 豆豉略為壓蓉，放碗中，加少許水開匀。
5. 燒熱鑊，下油約 1/2 湯匙，爆香蒜蓉、乾葱、豆豉，加入鮮魷炒匀，灒酒，炒片刻即加入西芹、青椒、紅燈籠椒和汁料炒匀即可上碟。

1. Rinse fresh squid, remove membrane and internal organs. Cut checkers pattern on the surface of squid.
2. Rinse celery and torn away woody fibers.
3. Rinse green pepper and red bell pepper, seed and cut into pieces. Rinse shallots, peel and pat with a chopper.
4. Crush fermented black beans briefly, put into a bowl, add some water and mix well.
5. Heat wok with 1/2 tbsp of oil, sauté minced garlic, shallots, fermented black beans, add squid and stir well. Drizzle wine, add celery, green peppers, red peppers and sauce and stir well. Serve.

入廚貼士 | Cooking Tips

- 鮮魷魚剞花要在魷魚的肚內的一面剞，剞十字時，刀和鮮魷魚要成斜角 45 度，先橫剞，再直剞。
- Cut checkers pattern on the belly side of squid. The chppoer should be in 45 degree angle with the squid. Cut horizontally first and then vertically.

20 分鐘
20 minutes

4 人
Serves 4

雪菜炒年糕

Stir-fried Pudding Cake with Snow Cabbage

材料 | Ingredients

水浸上海年糕 6 條	6 strips Shanghai pudding cakes
雪菜 4 棵	4 stalks snow cabbage
雞湯 2 杯	2 cups chicken broth
蒜頭 2 粒	2 cloves garlic

調味料 | Seasonings

糖 1 茶匙	1 tsp sugar
麻油 1/2 茶匙	1/2 tsp sesame oil
老抽 1/2 茶匙	1/2 tsp soy sauce

做法 | Method

1. 上海年糕瀝乾水分，切片。
2. 雪菜洗淨，榨乾水分，棄去根部，成細粒。
3. 蒜頭去衣，洗淨、剁蓉。
4. 燒熱鑊，下油約 1/2 湯匙，爆香蒜蓉，加入雪菜、上海年糕拌勻，再下調味料和雞湯煮至上海年糕變軟而湯汁將收乾即可盛起。

1. Drain Shanghai pudding cakes and slice.
2. Rinse snow cabbage, drain, cut roots and dice.
3. Peel garlic, rinse and chop.
4. Heat wok with 1/2 tbsp of oil, sauté minced garlic, add snow cabbage and Shanghai pudding cake. Then add seasonings and chicken stock and cook until softened and soup dries up. Serve.

入廚貼士 | Cooking Tips

- 雪菜要用水加少許鹽浸，才可去除雪菜的鹹味，大約浸 30 分鐘。
- Soak snow cabbage in salty water for 30 minutes to remove excess salty taste.

惹味醬炒通菜

Stir-fried Water Spinach with Flavored Paste

10 分鐘
10 minutes

4 人
Serves 4

材料 | Ingredients

通菜 500 克	500 g water spinach
米酒 2 茶匙	2 tsps rice wine

配料 | Side Ingredients

紅辣椒 1 隻	1 red chili
馬拉盞 2 湯匙	2 tbsps Malaysian shrimp paste
蒜蓉 3 茶匙	3 tsps minced garlic
XO 醬 3 茶匙	3 tsps XO paste

調味料 | Seasonings

糖 1 茶匙	1 tsp sugar
雞粉 1 茶匙	1 tsp chicken powder

做法 | Method

1. 通菜洗淨，摘去老菜，瀝乾水分。
2. 紅辣椒洗淨，去籽，切絲。
3. 燒熱鑊，下油約 1/2 湯匙，爆香配料，放入通菜，灒酒，加調味料，灑少許水，炒片刻即可上碟。

1. Rinse water spinach, remove old leaves and drain.
2. Rinse red chili, seed, cut into shreds.
3. Heat wok with 1/2 tbsp of oil, sauté side ingredients. Add water spinach, drizzle wine and add seasonings. Sprinkle with some water, mix well. Serve.

入廚貼士 | Cooking Tips

- 炒通菜不要蓋鍋蓋，否則菜會變黑。
- Do not cover the lid of wok when stir-frying water spinach, otherwise it will turn black.

10 分鐘
10 minutes

4 人
Serves 4

素翅炒蛋白
Stir-fried Artificial Shark Fin with Egg Whites

材料 | Ingredients

素翅 60 克
雞蛋白 4 隻
芥蘭梗 2 棵
上湯 6 湯匙

60 g artificial shark fin
4 egg whites
2 stems broccoli
6 tbsps chicken broth

調味料 | Seasonings

鹽 1/2 茶匙	1/2 tsp salt
薑汁 1/3 茶匙	1/3 tsp ginger juice
胡椒粉 1/4 茶匙	1/4 tsp pepper
薑 4 片（切絲）	4 slices ginger (shredded)
浙醋少許	Some red vinegar

做法 | Method

1. 素翅洗淨，燒沸一鍋水，加入素翅略燙片刻，瀝乾水分備用。
2. 雞蛋白打入大碗中，輕輕拂勻，不要打至起泡。
3. 芥蘭梗洗淨，切粒備用。
4. 燒熱鑊，下油約 1/2 湯匙，加入素翅、芥蘭梗拌勻，下調味料和上湯拌勻，盛起備用。
5. 再燒熱鑊，下油約 1 湯匙，加入雞蛋白略炒，將其他材料回鑊，炒勻上碟。

1. Rinse artificial shark fin, bring a pot of water to a boil and blanch artificial shark fin for a while, drain.
2. Beat egg whites in a large bowl, gently whisk, do not beat till bubbles form.
3. Rinse kale stems and dice.
4. Heat wok with 1/2 tbsp of oil, add artificial shark fin, broccoli stems and mix well. Add seasonings and broth, mix well, dish up and set aside.
5. Heat wok again with 1 tbsp of oil, add egg whites and stir-fry. Return other ingredients to wok, stir well and dish up.

入廚貼士 | Cooking Tips

- 炒雞蛋白要下比較多油才會滑。
- Add more oil when stir-frying egg whites to make the eggs smooth.

素菜竹笙扒菜膽

Stir-fried Vegetables with Bamboo Fungus

材料 | Ingredients

上海白菜 250 克
竹笙 20 條
冬菇 8 朵
蒜蓉 3 粒量
薑 3 片
雞湯 3/4 杯

250 g Shanghai cabbage
20 bamboo fungus
8 dried black mushrooms
3 cloves minced garlic
3 slices ginger
3/4 cup chicken broth

15 分鐘
15 minutes

4 人
Serves 4

調味料 | Seasonings

鹽 1/2 茶匙	1/2 tsp salt
糖 1/4 茶匙	1/4 tsp sugar

芡汁 | Thickening

蠔油 2 茶匙	2 tsps oyster sauce
生粉 1 茶匙	1 tsp cornstarch
水 4 湯匙	4 tbsps water

做法 | Method

1. 竹笙浸軟，剪去頭尾。冬菇洗淨，去蒂。把竹笙和冬菇汆水，瀝乾水分。
2. 上海白菜洗淨，摘去老葉，切成菜膽備用。
3. 燒熱鑊，下油約 1/2 湯匙，爆香蒜蓉，放入 1/2 份雞湯，焯熟菜膽盛起備用。
4. 再燒熱鑊，下油約 1/2 湯匙，爆香薑片，下竹笙和冬菇拌勻，加入餘下雞湯煮約 5 分鐘。加上芡汁煮至汁稍濃，排放在菜膽上即可。

1. Soak bamboo fungus, cut heads and tails. Rinse dried black mushrooms, remove stalks. Blanch bamboo fungus and mushrooms and drain.
2. Rinse Shanghai cabbage, remove old leaves and cut into sections.
3. Heat wok with 1/2 tbsp of oil, sauté minced garlic, add half of chicken broth, blanch vegetables and dish up.
4. Heat wok again with 1/2 tbsp of oil, sauté ginger slices, add bamboo fungus and mushrooms and mix well. Add remaining chicken broth to cook for about 5 minutes. Add thickening and cook until sauce is slightly thickened, dish up.

入廚貼士 | Cooking Tips

- 竹笙一定要汆水才沒有異味。
- Bamboo fungus must be blanched to remove unpleasant smell.

20 分鐘
20 minutes

4 人
Serves 4

炒素雜錦

Stir-fried Assorted Vegetables

材料 | Ingredients

雜豆 4 湯匙
豆腐乾 2 件
素雞 1 條
竹筍 1 個
蒜蓉 1 湯匙

4 tbsps assorted beans
2 pieces pressed beancurd
1 stick vegetarian chicken
1 bamboo shoot
1 tbsp minced garlic

調味料 | Seasonings

蠔油 2 湯匙	2 tbsps oyster sauce
糖 1 茶匙	1 tsp sugar
生抽 1 茶匙	1 tsp soy sauce
鹽 1/3 茶匙	1/3 tsp salt

芡汁 | Thickening

生粉 1 茶匙	1 tsp cornstarch
水 2 湯匙	2 tbsps water

做法 | Method

1. 豆腐乾、素雞洗淨，切粒。雜豆洗淨，瀝乾水分。
2. 竹筍洗淨，去外殼，切片後再切粒。
3. 燒熱鑊，下油約 1/2 湯匙，爆香蒜蓉，加入雜豆、素雞、豆腐乾粒和竹筍粒拌少，加入調味料炒勻，最後下芡汁勾芡。

1. Rinse pressed beancurd and vegetarian chicken and dice. Rinse assorted beans and drain.
2. Rinse bamboo shoots, remove woody outer shell, slice and dice.
3. Heat wok with 1/2 tbsp of oil, sauté minced garlic, add assorted beans, vegetarian chicken, pressed beancurd and bamboo shoot dice and stir well. Add seasonings and stir well, thicken with cornstarch solution.

入廚貼士 | Cooking Tips

- 蠔油可改為豆瓣醬，味道較辛辣，可更刺激味蕾。
- Oyster sauce could be replaced by chili bean paste in order to have spicy taste.

炒雜菜

Stir-fried Mixed Vegetables

20 分鐘
20 minutes

5 人
Serves 5

材料 | Ingredients

蓮藕 1/4 個	1/4 lotus roots
鮮百合 2 個	2 fresh lily bulbs
小籚筍 1 小扎	1 small bundle asparagus
紅辣椒絲 1 隻量	1 red chili (shredded)
蒜蓉 1 湯匙	1 tbsp minced garlic
米酒 2 茶匙	2 tsps rice wine
薑絲 1 茶匙	1 tsp shredded ginger

調味料 | Seasonings

魚露 2 茶匙	2 tsps fish sauce
糖 1/2 茶匙	1/2 tsp sugar

做法 | Method

1. 蓮藕洗淨，去皮，切薄片。
2. 百合洗淨，撕成一瓣瓣。
3. 小籚筍洗淨，切段。
4. 燒熱鑊，下油約 1/2 湯匙，爆香蒜蓉、薑絲和紅辣椒，先加入蓮藕炒熟，再下小籚筍炒片刻；加入百合和調味料料拌匀，灒酒，炒匀即可上碟。

1. Peel lotus root, rinse and cut into thin slices.
2. Rinse lily bulbs, separate.
3. Cut asparagus into sections.
4. Heat wok with 1/2 tbsp of oil, sauté minced garlic, ginger shreds and red chili. Add lotus root and stir-fry, add asparagus and cook for a while. Add lily bulbs and seasonings and mix well. Drizzle wine, stir well and dish up.

入廚貼士 | Cooking Tips

- 雜菜可轉用其他菜，如通菜、菠菜、西蘭花、莧菜等。
- Other mixed vegetables can be used instead, such as water spinach, spinach, broccoli, amaranth and so on.

10 分鐘
10 minutes

4 人
Serves 4

番茄炒蛋

Scrambled Eggs with Tomatoes

材料 | Ingredients

番茄 4 個	4 tomatoes
雞蛋 4 隻	4 eggs

調味料 | Seasonings

茄汁 2 湯匙	2 tbsps ketchup
糖 1 1/2 茶匙	1 1/2 tsp sugar
鹽 1/2 茶匙	1/2 tsp salt

做法 | Method

1. 番茄洗淨，切成 4 件。
2. 雞蛋打在碗中，加魚露打勻成蛋液。
3. 燒熱鑊，下油約 1/2 湯匙，下番茄炒，加調味料炒至番茄腍而不爛，盛起。
4. 再燒熱鑊，下油約 1 湯匙，倒下蛋液，快手炒至凝固，即將番茄回鑊，拌勻即可。

1. Rinse tomatoes, cut into 4 pieces.
2. Beat eggs in a bowl, add fish sauce and beat into egg mixture.
3. Heat wok with 1/2 tbsp of oil, add tomatoes and seasonings and stir-fry until cooked but not break, dish up.
4. Heat wok again with 1 tbsp of oil, add eggs, stir-fry until solidifies. Return tomatoes and mix well. Serve.

入廚貼士 | Cooking Tips

- 炒番茄時不用加水。
- Do not add water when stir-frying tomatoes.

蒜蓉椒絲炒生菜

Stir-fried Lettuce with Minced Garlic and Red Chili Shreds

10 分鐘
10 minutes

4 人
Serves 4

材料 | Ingredients

唐生菜 400 克 400 g Chinese lettuce
蒜頭 6 粒（剁蓉） 6 cloves garlic (minced)
紅辣椒 1 隻 1 red chili

調味料 | Seasonings

鹽 3/4 茶匙 3/4 tsp salt
糖 1/2 茶匙 1/2 tsp sugar

做法 | Method

1. 唐生菜洗淨，切去根部，每棵直切成兩半或四份備用。
2. 紅辣椒洗淨，去籽，切絲。
3. 燒熱鑊，下油約 1/2 湯匙，爆香蒜蓉和紅辣椒，下生菜拌勻，加入調味料拌勻即可上碟。

1. Rinse lettuce, cut roots and cut horizontally into halves or pieces.
2. Rinse red chili, seed and cut into shreds.
3. Heat wok with oil 1/2 tbsp of oil, sauté minced garlic and red chili, add lettuce and mix well. Add seasonings and mix well. Serve.

入廚貼士 | Cooking Tips

- 加入紅辣椒的餸菜會比較辣，如不喜歡太辣可不加。
- Red chili should not be added if you do not like spicy dish.

10 分鐘
10 minutes

4 人
Serves 4

腰果炒雞丁

Stir-fried Chicken Dice with Cashew Nuts

材料 | Ingredients

雞肉 250 克
腰果 100 克
青椒 1 個
紅燈籠椒 1 個
蒜蓉 1 茶匙
薑米 1 茶匙

250 g chicken
100 g cashew nuts
1 green pepper
1 red bell pepper
1 tsp minced garlic
1 tsp minced ginger

醃料 | Marinade

生粉 1 茶匙	1 tsp cornstarch
生抽 1 茶匙	1 tsp soy sauce
鹽 1/2 茶匙	1/2 tsp salt
雞粉 1/2 茶匙	1/2 tsp chicken powder

調味料 | Seasonings

鮮露 1 茶匙	1 tsp fresh fish sauce
生抽 1 茶匙	1 tsp soy sauce
生粉 1 茶匙	1 tsp cornstarch
水 5 湯匙	5 tbsps water

做法 | Method

1. 腰果洗淨，用水浸 10 分鐘後撈起，瀝乾水分。放入預熱至 150℃的焗爐內焗 20-25 分鐘。
2. 雞肉洗淨，切丁方塊，加醃料醃 20 分鐘。
3. 青椒和紅燈籠椒洗淨，去籽，切角。
4. 燒熱鑊，下油約 1/2 湯匙，爆香蒜蓉和薑米，下雞肉炒至 7 成熟，再下青椒和紅燈籠椒繼續炒至雞肉熟，加調味料拌勻，熄火，下腰果拌勻即可上碟。

1. Rinse cashew nuts, soak in water for10 minutes and drain. Bake in a preheated oven and bake at 150 ℃ for 20-25 minutes.
2. Rinse and dice chicken, marinate for 20 minutes.
3. Rinse green pepper and red bell pepper, seed and cut into wedges.
4. Heat wok with 1/2 tbsp of oil, sauté minced garlic and ginger, add chicken and stir-fry until 70% done. Then add green peppers and red bell pepper, stir-fry until chicken is done. Add seasonings and mix well, turn off heat, add cashew nuts and mix well. Serve.

入廚貼士 | Cooking Tips

- 炒腰果要熄火後才快速拌勻，效果才會脆。
- In order to keep the cripsy texture, cashew nuts should be added after turning off heat.

蜜味柚子炒雞球

Stir-fried Chicken in Pomelo Honey Sauce

材料 | Ingredients

雞髀 2 隻
紅燈籠椒 1 個

2 chicken thighs
1 red bell pepper

20 分鐘
20 minutes

4 人
Serves 4

醃料 | Marinade

蠔油 2 湯匙 — 2 tbsps oyster sauce
生粉 2 茶匙 — 2 tsps cornstarch
胡椒粉 1/2 茶匙 — 1/2 tsp pepper

調味料 | Seasonings

柚子蜜 4 湯匙 — 4 tbsps pomelo honey
蠔油 2 茶匙 — 2 tsps oyster sauce
生粉 1 茶匙 — 1 tsp cornstarch
魚露 1/2 茶匙 — 1/2 tsp fish sauce
水 6 湯匙 — 6 tbsps water

做法 | Method

1. 雞髀起肉，洗淨，切塊，瀝乾水分，加醃料拌勻，醃 30 分鐘。
2. 紅燈籠椒洗淨，去籽、切絲。
3. 燒熱鑊，下油約 1/2 湯匙，下雞塊煎至兩面金黃，盛起備用。
4. 燒熱鑊，下調味料，煮滾後加入紅椒燈籠椒絲拌炒勻，淋在雞塊上即可。

1. Remove bones from chicken thighs, rinse and cut into pieces, drain. Marinate for 30 minutes.
2. Rinse red bell pepper, seed and cut into shreds.
3. Heat wok with 1/2 tbsp of oil, pan-fry chicken until golden on both sides, dish up and set aside.
4. Heat wok, add seasonings and bring to a boil, add red pepper shreds and stir-fry. Pour onto chicken. Serve.

入廚貼士 | Cooking Tips

- 雞肉可改用其他肉類，如牛肉、豬肉或魚肉。
- Chicken meat could be replaced by beef, pork or fish.

25 分鐘
25 minutes

4 人
Serves 4

紫蘿雞片

Stir-fried Chicken with Pickled Ginger

材料 | Ingredients

子薑 160 克
罐頭菠蘿 1 罐（235 克）
雞髀 1 隻
蒜頭 4 粒
葱 3 棵

160 g young ginger
1 can canned pineapple (235 g)
1 chicken thigh
4 cloves garlic
3 stalks spring onion

醃料 | Marinade

生粉 1 茶匙	1 tsp cornstarch
鹽 1/2 茶匙	1/2 tsp salt
胡椒粉 1/2 茶匙	1/2 tsp pepper
糖 1/3 茶匙	1/3 tsp sugar

調味料 | Seasonings

生抽 1/2 茶匙	1/2 tsp soy sauce
麻油 1/2 茶匙	1/2 tsp sesame oil
生粉 1/2 茶匙	1/2 tsp cornstarch
糖 1/4 茶匙	1/4 tsp sugar
水 1 湯匙	1 tbsp water

做法 | Method

1. 雞髀洗淨，去骨，切片，以醃料醃 1/2 小時。
2. 菠蘿切件，糖水 1/2 杯留用。
3. 燒熱鑊，下油約 1/2 湯匙，爆香蒜頭，下雞片炒至熟。
4. 將糖水加入煮片刻，加入調味料、子薑和葱段拌匀，下菠蘿炒匀即可。

1. Rinse chicken thigh, remove bone and slice. Marinate for 1/2 hour.
2. Cut pineapple into pieces, retaine 1/2 cup of syrup.
3. Heat wok with 1/2 tbsp of oil, sauté garlic, stir-fry chicken pieces until done.
4. Cook syrup for a moment, add seasonings, young ginger and spring onion and mix well, add pineapple and stir well. Serve

入廚貼士 | Cooking Tips

- 菠蘿可以用新鮮的，但調味的糖要多加一些。
- Fresh pineapple can be used but some more sugar is needed.

醬爆雞球

Stir-fried Chicken Pieces in Soy Paste

材料 | Ingredients

雞髀 2 隻
乾葱頭 6 粒
葱 6 棵
磨豉醬 1 湯匙
薑米 2 茶匙

2 chicken thighs
6 cloves shallots
6 stalks spring onions
1 tbsp ground soy paste
2 tsps chopped ginger

20 分鐘
20 minutes

3 人
Serves 3

醃料 | Marinade

酒 1 湯匙	1 tbsp wine
薑汁 1 茶匙	1 tsp ginger juice
鹽 1/4 茶匙	1/4 tsp salt
胡椒粉 1/4 茶匙	1/4 tsp pepper

調味料 | Seasonings

生抽 1 湯匙	1 tbsp soy sauce
糖 1 茶匙	1 tsp sugar
麻油 1/2 茶匙	1/2 tsp sesame oil
水 1/2 杯	1/2 cup water

做法 | Method

1. 雞髀起肉，洗淨，切塊。用醃料醃 30 分鐘，備用。
2. 乾葱洗淨，去衣，拍扁。葱洗淨，切去根部和尾部，切度。
3. 燒熱鑊，下油約 1/2 湯匙，爆香乾葱、薑米和磨豉醬，加入雞塊和調味料炒至雞肉熟透即可。

1. Remove bones from chicken thigh, rinse and cut into pieces. Marinate for 30 minutes.
2. Rinse shallots, peel and pat. Rinse spring onions, Remove roots and tails, cut into sections.
3. Heat wok with 1/2 tbsp of oil, sauté shallots, minced ginger and ground soy paste, add chicken pieces and seasonings and stir-fry until done.

入廚貼士 | Cooking Tips

- 每種磨豉醬的鹹度也不同，所以一定要試味。
- Taste testing is needed as ground soy paste in different brands are in different salty level.

20 分鐘
20 minutes

4 人
Serves 4

三絲炒煙鴨胸

Stir-fried Smoked Duck Breast with Vegetable Shreds

材料 | Ingredients

煙鴨胸 1 件
西芹 2 條
青椒 1 個
紅蘿蔔 1/2 個
蒜頭 2 粒

1 piece smoked duck breast
2 stalks celery
1 green pepper
1/2 carrot
2 cloves garlic

調味料 | Seasonings

生抽 1/2 茶匙	1/2 tsp soy sauce
麻油 1/2 茶匙	1/2 tsp sesame oil
糖 1/3 茶匙	1/3 tsp sugar
生粉水 1 湯匙	1 tbsp cornstarch solution

做法 | Method

1. 蒜頭去衣，洗淨，拍扁。
2. 紅蘿蔔洗淨，去皮，切絲。西芹洗淨，撕去老根，切絲。青椒洗淨，去籽，切絲。
3. 燒熱鑊，下油約 1/2 湯匙，下鴨胸煎至兩面金黃，盛起。
4. 燒熱鑊，下油約 1/2 湯匙，爆香蒜頭，下紅蘿蔔絲略炒，再加入西芹和青椒絲，加調味料料拌勻，將煙鴨胸回鑊拌勻即成。

1. Peel garlic, rinse and pat.
2. Rinse carrot and peel, cut into shreds. Rinse celery, torn away woody fibers, shred. Rinse green peppers, seed and cut into shreds.
3. Heat wok with 1/2 tbsp of oil, pan-fry smoked duck breast until golden on both sides, dish up.
4. Heat wok with 1/2 tbsp of oil, sauté garlic, add carrot shreds, then add celery and green pepper. Add seasonings and mix well, return smoked duck breast to wok and mix well. Serve.

入廚貼士 | Cooking Tips

- 紅蘿蔔絲要先炒片刻才可加其他材料，因西芹和青椒都比較容易熟。
- Stir-fry carrot before adding other ingredients since celery and green peppers will be cooked in shorter time.

乾隆炒鴿鬆

Stir-fried Pigeon Meat

材料 | Ingredients

乳鴿 1 隻	1 pigeon
金華火腿 2 湯匙	2 tbsps Jinhua ham
馬蹄 12 粒	12 water chestnuts
冬菇 6 朵	6 dried black mushrooms
生菜 1 個	1 lettuce
甘筍 1/2 條	1/2 carrot
松子仁 2 湯匙	2 tbsps pine nuts
葱蓉 1 湯匙	1 tbsp minced shallots
蒜頭 1 粒	1 clove garlic
甜麵醬 1/2 杯	1/2 cup sweet bean paste
米酒 2 茶匙	2 tsps rice wine

醃料 | Marinade

雞蛋白 1 隻	1 egg white
生粉 1 茶匙	1 tsp cornstarch
鹽 1/2 茶匙	1/2 tsp salt
胡椒粉 1/4 茶匙	1/4 tsp pepper

30 分鐘
30 minutes

4 人
Serves 4

調味料 | Seasonings

糖 1 茶匙	1 tsp sugar
生抽 1 茶匙	1 tsp soy sauce
米酒 1 茶匙	1 tsp rice wine
麻油 1/2 茶匙	1/2 tsp sesame oil
胡椒粉 1/4 茶匙	1/4 tsp pepper
水 1 湯匙	1 tbsp water

做法 | Method

1. 材料洗淨，瀝乾。乳鴿起肉，切粒，下醃料醃 20 分鐘。
2. 燒熱鑊，下油約 1 碗，待油燒滾後下乳鴿泡油，盛起，瀝乾油分。
3. 馬蹄、甘筍去皮，切粒。生菜修剪成圓形。
4. 冬菇浸軟，去蒂，切粒。金華火腿汆水後切粒。
5. 燒熱鑊，下油約 1/2 湯匙，爆香蒜蓉，加入冬菇、馬蹄、甘筍、金華火腿粒拌炒，將乳鴿回鑊，灒酒，拌勻，灑下葱粒、松子仁拌勻，盛起。
6. 吃時用生菜包入餡料，沾上甜麵醬同食。

1. Rinse and drain ingredients. Remove bones from pigeon and dice, marinate for 20 minutes.
2. Heat wok, bring a bowl of oil to a boil, blanch pigeon, dish up and drain.
3. Peel and dice water chestnuts and carrot. Cut lettuce into circles.
4. Soak dried black mushrooms until soft, remove stalks and dice. Blanch and dice Jinhua ham.
5. Heat wok with 1/2 tbsp of oil, sauté minced garlic, add dried black mushrooms, water chestnuts, carrot, Jinhua ham and stir-fry well. Return pigeon to wok, drizzle wine, mix well. Sprinkle spring onion and pine nuts, mix well and dish up.
6. Serve with lettuce and sweet bean paste.

入廚貼士 | Cooking Tips

- 乳鴿可改用蠔豉、豬瘦肉或雞肉。
- Pigeon can be replaced by dried oysters, lean pork or chicken.

30 分鐘
30 minutes

4 人
Serves 4

洋葱炒鴨片

Stir-fried Onion with Duck Slices

材料 | Ingredients

鴨胸 2 大塊
葱 3 棵
洋葱 1 個
蒜蓉 1 湯匙
薑米 1 茶匙
米酒 2 茶匙

2 large pieces duck breast
3 stalks spring onion
1 onion
1 tbsp minced garlic
1 tsp minced ginger
2 tsps rice wine

調味料 | Seasonings

生粉 1 茶匙	1 tsp cornstarch
鹽 1/2 茶匙	1/2 tsp salt
生抽 1/2 茶匙	1/2 tsp soy sauce
胡椒粉 1/2 茶匙	1/2 tsp pepper

調味料 | Seasonings

老抽 1/2 茶匙	1/2 tsp dark soy sauce
生抽 1/2 茶匙	1/2 tsp soy sauce
糖 1/3 茶匙	1/3 tsp sugar
生粉水 1 湯匙	1 tbsp cornstarch solution

做法 | Method

1. 材料洗淨，瀝乾。鴨胸切片，下醃料醃 30 分鐘。
2. 洋葱去衣，切絲。葱切段。
3. 燒熱鑊，下油約 1/2 湯匙，爆香洋葱，加少許鹽炒至微黃色，盛起待用。
4. 再燒熱鑊，下油約 1/2 湯匙，爆香蒜蓉和薑米，下鴨片炒至熟，將洋葱回鑊，加入調味料和葱段拌勻，灒酒即可上碟。

1. Rinse and drain ingredients. Rinse duck breast, slice, marinate for 30 minutes.
2. Peel and shred onion. Cut spring onion into sections.
3. Heat wok with 1/2 tbsp of oil, sauté onion, add some salt and stir-fry until slightly brown, dish up.
4. Heat wok again, add 1/2 tbsp of oil and sauté minced garlic and ginger. Stir-fry duck pieces until done. Return onion to wok, add seasonings and spring onion and mix well, drizzle wine. Serve.

入廚貼士 | Cooking Tips

- 如果用全隻鴨，不要忘記除去鴨尾的白色子。
- Remember to remove the white piece in duck tail if whole duck is used.

15 分鐘
15 minutes

4 人
Serves 4

荷豆炒臘腸

Stir-fried Chinese Sausage with Snow Peas

材料 | Ingredients

臘腸 1 孖
荷蘭豆 120 克
芹菜 50 克
薑 4 片
蒜蓉 2 粒量
米酒 1 茶匙

1 pair Chinese sausage
120 g snow peas
50 g Chinese celery
4 ginger slices
2 cloves garlic (minced)
1 tsp rice wine

汁料 | Sauce

上湯 3 湯匙	3 tbsps chicken broth
糖 1/3 茶匙	1/3 tsp sugar
鹽 1/3 茶匙	1/3 tsp salt
生粉 1 茶匙	1 tsp cornstarch

做法 | Method

1. 臘腸放溫水中沖洗，取出置碟中。
2. 臘腸用大火隔水蒸約 10 分鐘，取出，待涼後切片。
3. 荷蘭豆洗淨，撕去老根。芹菜洗淨，切去根部，切段。
4. 燒熱鑊，下油約 1/2 湯匙，爆香蒜蓉和薑片，下荷蘭豆和臘腸，加水 2 湯匙，加蓋焗 1 分鐘，灒酒，放芹菜和汁料，炒勻即可上碟。

1. Rinse Chinese sausages in warm water, remove and put onto a plate.
2. Steam Chinese sausages over high heat for 10 minutes, leave to cool and slice.
3. Rinse snow peas, torn away woody fibres. Rinse Chinese celery, remove roots and section.
4. Heat wok with 1/2 tbsp of oil, sauté minced garlic and ginger slices, add snow peas and Chinese sausage. Add 2 tbsps of water, cover and cook for 1 minute, drizzle wine. Add Chinese celery and sauce, then stir well. Serve.

入廚貼士 | Cooking Tips

- 臘腸先要用溫水洗去表面的灰塵。臘腸蒸後，比較容易切片。
- Chinese sausages should be rinsed in warm water to remove dust on surface. It could be easily sliced after steaming.

生筋肉碎炒芹菜

Stir-fried Chinese Celery with Deep-fried Beanburd Balls and Minced Pork

材料 | Ingredients

絞豬肉 150 克
生筋 6 個
芹菜 2 棵

150 g minced pork
6 deep-fried beanburd balls
2 stalks Chinese celery

30 分鐘
30 minutes

4 人
Serves 4

醃料 | Marinade

鹽 1/2 茶匙	1/2 tsp salt
生抽 1/2 茶匙	1/2 tsp soy sauce
胡椒粉 1/2 茶匙	1/2 tsp pepper

調味料 | Seasonings

鹽 1/2 茶匙	1/2 tsp salt
雞粉 1/2 茶匙	1/2 tsp chicken powder
麻油 1/2 茶匙	1/2 tsp sesame oil

做法 | Method

1. 絞豬肉加醃料醃 20 分鐘。
2. 生筋洗淨，汆水。
3. 芹菜洗淨，切去根部，去葉，洗淨後切段。
4. 燒熱鑊，下油約 1/2 湯匙，爆香絞豬肉，炒至 7 成熟，加入生筋和芹菜拌勻，下調味料拌勻即可。

1. Marinate minced pork for 20 minutes.
2. Rinse deep-fried beancurd balls and blanch.
3. Rinse Chinese celery, remove woody fibers and leaves. Rinse and cut into sections.
4. Heat wok with 1/2 tbsp of oil, sauté minced pork, stir-fry till 70% done. Add deep-fried beancurd balls and Chinese celery, add seasonings and mix well.

入廚貼士 | Cooking Tips

- 生筋汆水是為了去除油溢味。
- Blanching deep-fried beancurd balls could remove excess oil smell.

25 分鐘
25 minutes

5 人
Serves 5

炒雜錦粒粒
Stir-fried Assorted Dice

材料 | Ingredients

雞肉粒 150 克	150 g chicken dice
冬菇 4 朵	4 dried black mushrooms
五香豆腐乾 3 件	3 pieses spiced pressed beancurd
菜脯粒 2 湯匙	2 tbsps perserved radish dice
葱粒 2 棵	2 stalks spring onion (dice)
紅辣椒 1 隻	1 red chili
洋葱粒 1/2 個	1/2 onion (dice)
蝦米 1/4 杯	1/4 cup dried shrimps

調味料 | Seasonings

蒜蓉 2 茶匙	2 tsps minced garlic
豆瓣醬 1 茶匙	1 tsp broad bean paste
鹽 1/4 茶匙	1/4 tsp salt

做法 | Method

1. 五香豆腐乾洗淨，切粒。
2. 燒熱鑊，下油約 1/2 湯匙，爆香洋葱，盛起備用。
3. 再燒熱鑊，下油約 1/2 湯匙，爆香蝦米、冬菇，盛起備用。
4. 再燒熱鑊，下油約 1/2 湯匙，爆香蒜蓉、豆瓣醬、紅辣椒，下雞肉炒至 7 成熟，將其他材料回鑊，拌炒勻即可。

1. Rinse spiced pressed beancurd and dice.
2. Heat wok with 1/2 tbsp of oil, sauté onion, dish up.
3. Heat wok again with 1/2 tbsp of oil, sauté dried shrimps, dried black mushrooms, dish up and set aside.
4. Heat wok again with 1/2 tbsp of oil, sauté minced garlic, broad bean paste, red chili, add chicken and stir-fry until 70% done. Return other ingredients to wok, stir-fry and mix well. Serve.

入廚貼士 | Cooking Tips

- 菜脯可用雪菜代替，其他材料也可隨個人口味更改。
- Preserved radish could be replaced by pickled vegetables, other ingredients can also be changed according to personal taste.

回鍋肉

Double-cooked Pork

材料 | Ingredients

豬脖肉 300 克
紹菜 300 克
豆腐乾 3 塊
蒜頭 2 粒
紅燈籠椒 1 隻
青椒 1 隻

300 g heel meat of pig
300 g cabbage
3 pieces pressed beancurd
2 cloves garlic
1 red bell pepper
1 green pepper

25 分鐘
25 minutes

4 人
Serves 4

調味料 | Seasonings

甜麵醬 2 湯匙	2 tbsps sweet bean paste
辣豆瓣醬 2 茶匙	2 tsps spicy broad bean paste

汁料 | Sauce

生抽 2 茶匙	2 tsps soy sauce
鹽 1/2 茶匙	1/2 tsp salt
糖 1/2 茶匙	1/2 tsp sugar
老抽 1/2 茶匙	1/2 tsp dark soy sauce

做法 | Method

1. 材料洗淨，瀝乾水分。燒滾一鍋水，下豬踭肉，用中火煮 1/2 小時，撈起，瀝乾，待涼後切薄片。
2. 豆腐乾切片。蒜頭去衣，切片。
3. 青椒、紅燈籠椒去籽，切塊。紹菜切塊。
4. 燒熱鑊，下油約 1/2 湯匙，爆香青椒、紅燈籠椒，盛起。
5. 再燒熱鑊，下油約 1/2 湯匙，爆香蒜片，加入青椒、紅燈籠椒、豆腐乾、豬肉片和調味料炒勻，倒入汁料，灒酒，加入紹菜兜勻後即可上碟。

1. Rinse and drain ingredients. Bring a pot of water to a boil, add heel meat of pig and simmer for 1/2 hours over medium heat. Dish up and drain, leave to cool and cut into thin slices.
2. Slice pressed beancurd. Peel and slice garlic.
3. Seed and cut green peppers and red bell pepper into pieces. Cut cabbage into pieces.
4. Heat wok with 1/2 tbsp of oil, sauté green pepper and red bell pepper, dish up.
5. Heat wok again with 1/2 tbsp of oil, sauté garlic slices, add green pepper, red bell pepper, pressed beancurd, pork slices and seasonings, stir well. Add sauce, drizzle wine and add cabbage. Serve.

入廚貼士 | Cooking Tips

- 豬肉要放雪櫃雪凍才可以切成薄片。
- Chill pork in refrigerator before cutting into thin slices.

15 分鐘
15 minutes

4 人
Serves 4

絲瓜炒豬膶

Stir-fried Loofah with Pig's Liver

材料 | Ingredients

豬膶 320 克	320 g pig's liver
絲瓜 1 條	1 loofah
紹酒 1 湯匙	1 tbsp Shaoxing wine
薑絲 1 茶匙	1 tsp shredded ginger

醃料 | Marinade

薑汁酒 1 湯匙	1 tbsp ginger wine
生粉水 1/2 湯匙	1/2 tbsp cornstarch solution

調味料 | Seasonings

生抽 1/2 湯匙	1/2 tbsp soy sauce
糖 1/2 茶匙	1/2 tsp sugar
鹽 1/4 茶匙	1/4 tsp salt

做法 | Method

1. 將豬膶切厚片，用醃料稍醃。燒滾一鍋水，豬膶放滾水中略燙，撈起。
2. 絲瓜去皮，洗淨，切塊。
3. 鑊燒油熱，爆香薑絲，放入絲瓜和酒炒透，再加入豬膶及調味料，快手炒片刻，即可。

1. Cut pig's liver into thick slices and marinate. Bring a wok of water to a boil, blanch pig's liver for a while and dish up.
2. Peel loofah, rinse and cut into pieces.
3. Heat wok with oil, sauté ginger shreds, add loofah and wine, stir-fry. Add pig's liver and seasonings, stir-fry for a while. Serve.

入廚貼士 | Cooking Tips

- 豬膶先浸泡在稀釋的醋水中，再沖洗乾淨，可去除腥味。
- Pig's liver can be soaked in diluted vinegar water and rinse to remove unpleasant smell.

菜芯炒牛肉

Stir-fried Beef with Choi Sum

材料 | Ingredients

菜芯 300 克	300 g Choi Sum
牛肉 150 克	150 g beef
蒜蓉 1 湯匙	1 tbsp minced garlic

醃料 | Marinade

生抽 1 茶匙	1 tsp soy sauce
生油 1 茶匙（後下）	1 tsp oil (add later)
生粉 1 茶匙	1 tsp cornstarch
糖 1/3 茶匙	1/3 tsp sugar
水 1 湯匙	1 tbsp water

10 分鐘
10 minutes

3 人
Serves 3

調味料 | Seasonings

鹽 1/2 茶匙　　1/2 tsp salt
雞粉 1/2 茶匙　　1/2 tsp chicken powder

做法 | Method

1. 牛肉洗淨，切片，下醃料拌勻，以油蓋面，待 20 分鐘。
2. 菜芯洗淨，切度。
3. 燒熱鑊，下油約 1/2 湯匙，下牛肉快手兜勻至 7 成熟，盛起。
4. 再燒熱鑊，下油約 1/2 湯匙，爆香蒜蓉，下菜芯和調味料炒至 7 成熟，將牛肉回鑊，炒勻即可上碟。

1. Rinse beef and slice, marinate and mix well. Cover the surface with oil, leave for 20 minutes.
2. Rinse Choi Sum and cut into sections.
3. Heat wok with 1/2 tbsp of oil, add beef and stir-fry until 70% done.
4. Heat wok again with 1/2 tbsp of oil, sauté minced garlic, Choi Sum and seasonings and stir-fry until 70% done. Return beef to wok, stir well and dish up.

入廚貼士 | Cooking Tips

- 醃牛肉時要先下其他醃料拌勻，再逐少加水，拌勻後再加，最後加油蓋着。
- When marinating beef, add other ingredients and mix well, then gradually add water, mix well. Finally add oil to cover beef.

20 分鐘
20 minutes

4 人
Serves 4

豉汁涼瓜炒牛肉

Stir-fried Beef with Bitter Gourd in Black Bean Sauce

材料 | Ingredients

牛肉 150 克	150 g beef
涼瓜 2 個	2 bitter gourds
豆豉 1 湯匙	1 tbsp fermented black beans
蒜蓉 1 湯匙	1 tbsp minced garlic

牛肉醃料 | Marinade for beef

生抽 1 茶匙	1 tsp soy sauce
生油 1 茶匙（後下）	1 tsp oil (addlater)
生粉 1 茶匙	1 tsp cornstarch
糖 1/3 茶匙	1/3 tsp sugar
水 1 湯匙	1 tbsp water

涼瓜醃料 | Marinade for bitter gourd

糖 2 湯匙　　2 tbsps sugar
鹽 1/2 茶匙　　1/2 tsp salt

調味料 | Seasonings

鹽 1/2 茶匙　　1/2 tsp salt

做法 | Method

1. 牛肉洗淨，切片，下牛肉醃料拌勻，以油蓋面，待 20 分鐘。
2. 涼瓜洗淨，去囊，切片，加涼瓜醃料醃 10 分鐘。
3. 豆豉壓扁，放碗中，用少許水開好，備用。
4. 燒熱鑊，下油約 1/2 湯匙，加入牛肉快手兜勻至 7 成熟即盛起。
5. 再燒熱鑊，下油約 1/2 湯匙，爆香蒜蓉和豆豉，下涼瓜拌炒，加鹽調味，加適量水。蓋好鑊蓋煮片刻，將牛肉回鑊即可。

1. Rinse beef and slice. Marinate beef and cover with oil, leave for 20 minutes.
2. Rinse bitter gourd and remove seeds, marinate for 10 minutes.
3. Pat fermented black beans, put in a bowl with some water, mix well.
4. Heat wok again with 1/2 tbsp of oil, add beef and stir-fry quickly until 70% done, dish up.
5. Heat wok again, add 1/2 tbsp of oil, sauté minced garlic and fermented black beans, then add bitter gourd and stir-fry well. Season with salt, add water, cook for a while with lid covered. Return beef to wok. Serve.

入廚貼士 | Cooking Tips

- 涼瓜先用鹽醃可令其變軟，再洗去鹽水。加入少許糖，可令涼瓜減去大部分苦味。
- Marinate bitter gourd with salt can make it soft, then rinse away salt. Add some sugar could reduce the bitterness.

香蒜牛肉粒

Stir-fried Beef Dice with Garlic

15 分鐘
15 minutes

4 人
Serves 4

材料 | Ingredients

牛肉 500 克	500 g beef
炸蒜蓉 2 湯匙	2 tbsps deep-fried garlic
紅辣椒 1 隻	1 red chili
蒜頭 1/2 個	1/2 clove garlic

調味料 | Seasonings

淮鹽 1/2 茶匙	1/2 tsp pepper salt

做法 | Method

1. 蒜頭去衣，洗淨，剁蓉。紅辣椒洗淨，去籽，切幼粒。
2. 牛肉洗淨，切成丁方。
3. 燒熱鑊，下油約 1/2 湯匙，改用中火，爆香蒜蓉，放入牛肉粒煎至 6 成熟，加入炸蒜蓉拌勻，加紅辣椒粒和淮鹽拌勻即可上碟。

1. Peel garlic, rinse and mince. Rinse red chili, seed and cut finely.
2. Rinse beef and dice.
3. Heat wok with 1/2 tbsp of oil, turn to medium heat, sauté minced garlic, pan-fry beef until 60% done. Add deep-fried garlic and mix well. Then add red pepper and pepper salt and mix well. Serve.

入廚貼士 | Cooking Tips

- 牛肉只可煎至 6 至 7 成熟，才不會太韌。
- To keep the soft texture, pan-fry beef until only 60-70% done.

編著
梁燕

編輯
紫彤

美術設計
Carol Fung

排版
何秋雲

翻譯
Faye Kwok

攝影
Fanny

出版者
萬里機構出版有限公司
香港鰂魚涌英皇道1065號東達中心1305室
電話：2564 7511
傳真：2565 5539
電郵：info@wanlibk.com
網址：http://www.wanlibk.com
http://www.facebook.com/wanlibk

發行者
香港聯合書刊物流有限公司
香港新界大埔汀麗路36號
中華商務印刷大廈3字樓
電話：2150 2100
傳真：2407 3062
電郵：info@suplogistics.com.hk

承印者
美雅印刷製本有限公司

出版日期
二零一八年十一月第一次印刷

Published in Hong Kong.
ISBN 978-962-14-6869-7